Creepy Crawlies

Wendy Einstein
& Einstein Sisters

KidsWorld

Bugs are everywhere!

Ugh! They **invade** our homes (and sometimes our bodies!) and skitter through our nightmares.

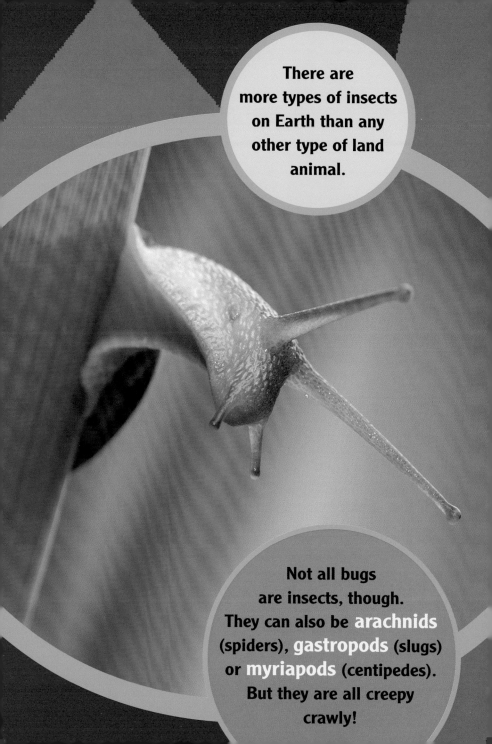

There are
more types of insects
on Earth than any
other type of land
animal.

Not all bugs
are insects, though.
They can also be **arachnids**
(spiders), **gastropods** (slugs)
or **myriapods** (centipedes).
But they are all creepy
crawly!

Earwig

Earwigs get their name from an old myth. People once believed these creepy creatures would crawl into a person's ear and lay their eggs on the person's brain.

Earwigs live on every continent except Antarctica. There are about 1800 species worldwide.

Usually you'll find them crawling on plants or in dark, damp places. But some species can fly!!!! Ack!!!!

They use the **huge pincers** on their butts to defend themselves if they are attacked.

Female earwigs are good moms. They lay their eggs in damp places to keep them moist and turn them over often so they don't get mouldy. They also protect them from predators, like centipedes and harvestmen. Once the eggs hatch, the mom stays with her young until they can take care of themselves.

Silverfish

Silverfish love dark, damp places and are especially attracted to things like cardboard boxes, books and any type of cloth. So think twice before you throw your damp towel on the floor next time you shower!

It is easy to see how silverfish got their name. Their bodies are covered with **tiny scales**, much like fish. They are also a shiny silver colour, like fish, and they move with a wriggling, side-to-side motion, a lot like how a fish swims.

Scientists have found fossilized silverfish that date back to the **Paleozoic Era**, more than 400 million years ago. That's more than 170 million years before the dinosaurs lived!

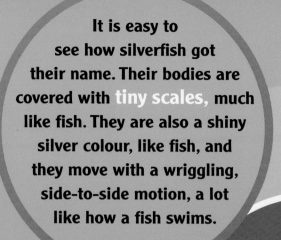

The praying mantis might look like a peaceful creature as it holds its arms folded as if in prayer, but don't be fooled. This bug is a killer!

Praying Mantis

The praying mantis is an ambush predator. It waits for its prey to get close and then strikes, stabbing the victim with one of its long, pointy front legs.

It can hide in plain sight because when it stays really still, it is camouflaged and looks like part of a plant.

This amazing insect eats other bugs as well as lizards, snakes, frogs, mice and even small birds, including hummingbirds and warblers. When it catches a bug, it bites the head off and eats that first.

Stick Insect

Another master of disguise is the **stick insect**. It looks just like a stick! Some species even sway gently back and forth to look like a twig blowing in the wind.

There are about 3000 species of stick insect in the world. They are most common in tropical forests.

Some species are HUGE!

Stick insects use **camouflage** to hide from predators, not to catch prey. They are **herbivores** and eat only plants.

Ever notice a black worm-like creature on your body after you've been swimming in a lake? That would be a leech, and it is drinking your blood!

You won't feel its bite because special chemicals in its saliva block the pain and keep the blood flowing. Leech bites can keep bleeding for 10 to 12 hours after the leech lets go.

Leech

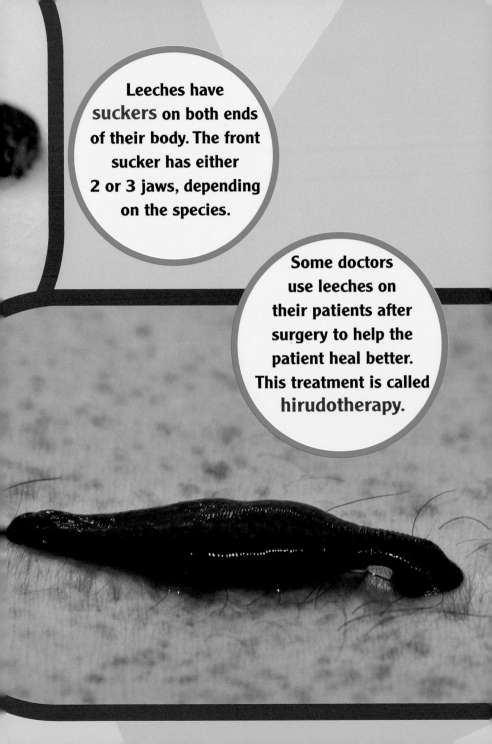

Leeches have **suckers** on both ends of their body. The front sucker has either 2 or 3 jaws, depending on the species.

Some doctors use leeches on their patients after surgery to help the patient heal better. This treatment is called **hirudotherapy**.

Camel Spider

The **camel spider**, also called the **wind scorpion**, is not a spider or a scorpion. It is a **solifuge**, which is a cousin of the spider and scorpion.

Most camel spiders live in deserts in the Middle East, but about 20 species can be found in North America.

These creatures do not use poison to kill their prey. Instead they use their huge jaws to chew their victim to death.

Camel spiders are fast! They can run almost as fast as a squirrel!

Bed-Bug

At night as you sleep peacefully in your bed, bed bugs are on the move. They crawl over your blankets or sheets looking for exposed skin. And then they bite and drink your blood!

Bed bugs have light-coloured bodies before they eat. After their blood meal, they are reddish-brown.

You won't usually see bed bugs. Instead, you'll find little spots of blood and maybe some bed bug poo on your sheets. And you'll have itchy welts on your skin.

These **creepy crawlies** are small and flat, about the size of the eraser on the end of a pencil.

During the day they hide in small spaces, like behind baseboards or in cracks in the walls or bedframe.

You may not be happy to see a house centipede speeding across your bathroom floor, but it is actually helpful to have around. It preys on the OTHER bugs you might not have known were living in your home.

House centipedes eat troublesome bugs like silverfish, bed bugs, ants, termites and cockroaches.

House Centipede

This centipede has 15 pairs of legs. The two front legs, close to the mouth, are called **forcipules**. They are full of **venom** the centipede injects into its prey.

To catch its prey, a house centipede jumps on its victim and wraps its long body around the doomed creature, kind of like a tiny, many-legged boa constrictor.

Coconut Crab

The coconut crab is one scary-looking creature. It is the biggest land crab in the world.

This giant crab has no natural predators other than other coconut crabs and humans.

As you might have guessed from its name, this crab often eats coconuts. The crab uses its super-strong **pinchers** to crush the hard coconut shell.

It will also eat rats, other coconut crabs, birds, carrion (dead animals) and fruit.

Coconut crabs can even climb trees!

Weta

Wetas are flightless cousins of grasshoppers and crickets. They live only in New Zealand.

The more than 70 species of weta are grouped into 5 different categories: ground, tree, cave, tusked and giant.

The giant weta is the heaviest insect in the world! It is as big as a hamster.

Some species of wetas are endangered and others are at risk because they are killed by rats and cats. Rats and cats are not part of natural ecosystem in New Zealand. They were brought to the island by settlers from other countries. Wetas have no defense against these predators.

Slug

A slug is basically a snail that has no shell. Like a snail, a slug moves around by sliding on a trail of its own slime.

The slug makes slime, called mucus, from its foot. The foot is always oozing mucus, kind of like how your nose drips snot when you have a cold. The mucus makes the slug slippery.

A slug has 2 sets of tentacles. The top set are called eye stalks and have eyes at the tip. They are also used for smelling. The bottom set are used for tasting and touch.

Being slimy can help protect the slug from predators. Slippery slugs are hard to grab, and many predators don't want to eat slimy food.

Slugs are hermaphrodites, meaning that every slug is both male and female.

Bull Ant

Bull ants are the largest ants in Australia and one of the largest ants in the world.

There are more than 90 species. They live in only in Australia, except one species that also lives in New Caladonia, a nearby island.

Bull ants have huge jaws and a stinger on their butt filled with venom. They grip their victim with their jaws and then bend their butt around to sting the victim. They can sting as many times as they want to.

These ants are also called bulldog ants.

Earthworm

Earthworms spend most of their time underground.

An earthworm's body is made up of many sections, called segments. If the worm loses a segment at the tail end of its body, it can regrow the segment.

An earthworm moves by pushing its front end forward until its whole body is stretched out. Then the worm pulls its tail end forward until its body is its normal length again.

Worm bodies may look smooth, but they are covered with tiny stiff hairs called setae. The setae help the worm move by gripping the ground as the worm pushes itself forward.

Worms don't have eyes. Instead, they have special cells on their skin that can sense light. They also don't have ears or a nose.

Maggot

Those little white worm-like creatures wriggling around on your rotten garbage are most likely maggots.

Maggots are fly larvae, young flies that just hatched.

Female flies lay their eggs on a food source so their young will have plenty to eat once they hatch. For flies, good food choices include poo, dead creatures, rotting food or garbage.

Maggots may look gross, but some doctors use them to clean wounds. The maggots are put into the wound and left there for 2 or 3 days. They eat the dead flesh in the wound, so only healthy tissue is left behind. This helps the wound heal faster with less chance of becoming infected.

Dung Beetle

And the award
for the strongest
creature in the world goes
to…the dung beetle! A dung
beetle can push a ball of
poo that is more than
1000 times heavier
than itself.

There are 3 main types of dung beetle. **Rollers** make big poo balls and roll them to a safe place to bury them for later.

Dung beetles eat poo! They also lay their eggs in it so their young can eat it when they hatch.

Tunnellers dig through a poo pile and live under it.

Dwellers live in the poo pile.

Dung beetles live on every continent except Antarctica. They don't like extreme cold, so you also won't find them in the far north in North America, Europe or Asia.

Flea

Fleas are tiny, about the size of the tip of a pen. They have no wings but have special legs that make them great jumpers.

Fleas can jump 200 times their own body length. If you could jump like a flea, you could easily leap from the ground onto the top the Eiffel Tower.

These little **bloodsuckers** usually live on furry mammals, like cats or dogs. They hide in the fur and bite their victim to drink its blood. Hungry fleas will also feed on people.

Fleas can live for up to 100 days without food. They often hide in carpets or furniture waiting for their next mammal victim to pass by.

Harvestman

Harvestmen might look like spiders, but they are not. Although they are related to spiders, they are actually more closely related to scorpions.

These creepy crawlies are also known as daddy longlegs.

Harvestmen cannot make silk, so they do not make webs. Some species make a special kind of glue to trap their prey.

When being attacked by a predator, the harvestman can drop a leg off its body. The dropped leg keeps twitching for a while, which distracts the would-be predator. Harvestmen cannot regrow legs they have lost.

If you could see what is lurking in your carpet, pillow, mattress, curtains and on flat surfaces in your home you would be horrified. These places are crawling with dust mites.

Dust Mite

These mites are related to spiders. They eat dust, which is made up of dead skin cells and dirt.

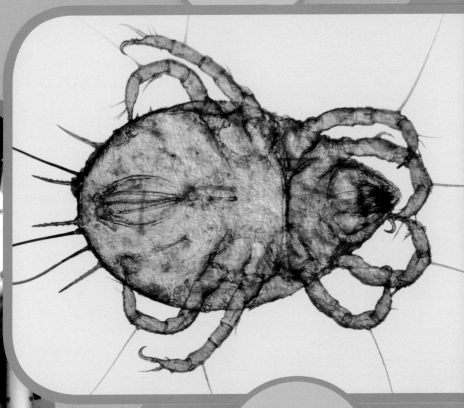

**Dust mites
are so small they
cannot be seen
without
a microscope.**

**Dust mites
are not dangerous
to people, but
they can cause
allergies.**

Treehopper

Some treehoppers look like thorns on a branch. This camouflage helps keep them safe from predators.

Because of this camouflage, treehoppers are also called thorn bugs.

These cutie crawlies (they are too cute to be called creepy) bite into tree branches to drink the sap.

There are more than 3000 known species of treehopper, but scientists think there may be thousands more that have not yet been identified.

Some species form **mutualistic** relationships with ants. The ants protect the treehoppers, and the treehoppers make honeydew, a sweet liquid the ants eat.

Centipede

Centipedes are the cheetahs of the bug world. They are super-fast runners, which must be tricky with all those feet.

Centipede means 100 feet, but most species don't have that many. Most types of centipede have between 30 and 50 feet, but a few species can have up to 300.

Centipedes are fierce predators. They chase down their prey and hold it with their legs as they give it a venomous bite.

Their front legs are specially adapted to be more like fangs than feet.

Millipede

Millipedes are like the centipede's slower, gentler cousins. They are herbivores, not predators, which means they eat only plants.

Scientists think millipedes may have been the first land creatures. Fossils show that millipedes lived more than 425 million years ago. That's more than 200 million years before the dinosaurs!

Most millipede species have between 40 and 350 legs, not 1000 as their name suggests (millipede means 1000 legs). One species that lives in the United States, *Illacme plenipes*, has 750 legs!

Millipedes roll into a ball when they feel threatened. Some species can release a smelly liquid to prevent predators from biting them.

Giraffe
Weevil

Male
Weevil

Another bug
that is more cute
then creepy is the
giraffe weevil. I'm sure
you can guess how it
got its name.

Weevils are
a type of beetle.
They live only on the
island of Madagascar
off the coast
of Africa.

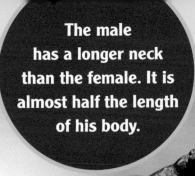

The male has a longer neck than the female. It is almost half the length of his body.

Female Weevil

Males fight each other to get a mate. They beat each other with their long neck and try to push the other weevil off the tree.

These beetles make one type of tree their home, the giraffe beetle tree.

Locust

A locust might not look too creepy, but they can be a real nightmare.

Locusts can eat their body weight of food every day. They feed on plants like grasses and grains.

These insects are usually solitary, meaning they live alone. But when there isn't enough food, locusts swarm together. Swarms can have more than 1000 million locusts and can cover hundreds of kilometers.

Swarming locusts can eat every plant in their path. If they can't find enough plants, they will even eat each other.

Botfly Larva

Have you been to South or Central America lately? Did you feel a weird lump on your skin? It could be a **botfly larva**.

Botfly larvea are **parasites**. Depending on the species, the larvae live on their host's skin or in the stomach.

The female human botfly catches a mosquito and lays her eggs on its abdomen. When the mosquito lands on a person, the eggs hatch and the larvae burrow under the person's skin.

The hole they make in the skin stays open so the larvae can breathe. The larvae also stick their butt out of the hole to poo.

It is hard to get the larvae out of your skin. Their bodies have sharp spines that point upward and dig into your skin if you try to pull the larvae out.

The larvae leave on their own when they are big enough (after 2 to 3 months). They wriggle out of the hole and drop to the ground, ready to turn into flies.

The **cockroach** is another creature that has been around since before the time of the dinosaurs.

There are more than 4500 species of cockroach in the world. Only a few species make pests of themselves by invading our homes and contaminating our food.

Cockroach

A cockroach can live without its head for about two weeks! The body can even still move around! Because it breathes through holes in its side and not its mouth, the roach doesn't need its head to breathe. It does need a mouth to eat though, so a headless roach will eventually starve to death.

Cockroaches will eat almost anything, even each other.

Head Louse

Head lice are only about the size of a sesame seed, but they can sure make your life miserable!

Head lice live in people's hair. They bite the person's scalp to drink blood. These lice can live for only about 24 hours off a person.

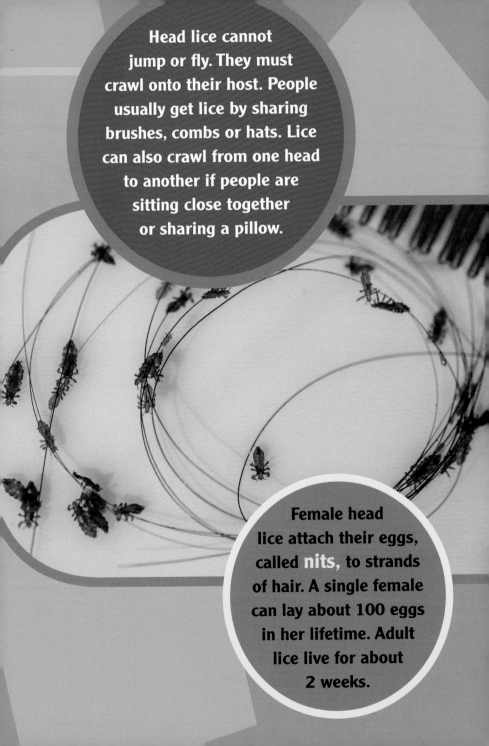

Head lice cannot jump or fly. They must crawl onto their host. People usually get lice by sharing brushes, combs or hats. Lice can also crawl from one head to another if people are sitting close together or sharing a pillow.

Female head lice attach their eggs, called nits, to strands of hair. A single female can lay about 100 eggs in her lifetime. Adult lice live for about 2 weeks.

Tick

Ticks are not insects; they are **arachnids**, the same family that spiders belong to. Like their spider cousins, ticks have 8 legs.

Ticks are parasites. They drink the blood of other animals to survive. Some species of tick, like the cattle tick, feed only on specific animals. Other types of ticks are less picky and feed on whatever creature they can grab on to.

When a tick drinks blood, its body swells like a balloon. A well-fed tick can expand to twice its normal size.

Many species of tick have no eyes. They have special sensors in their front legs that tell them when an animal is nearby.

Army Ant

Army ants
are always on the
move. They can form
living bridges by hang-
ing on to each other to
cross over gaps or
other obstacles.

These ants
live in tropical
rainforests in
Africa and South
and Central
America.

They are called army ants because they hunt in large groups, called **raids,** attacking any creature in their path. Because there are so many ants in a raid, they can kill creatures much larger than themselves.

When a raid is on the move, any creatures in its path try to get out of the way. Some animals, especially birds, take advantage of this and stay just out of reach, catching any creatures that try to escape the raid.

Scorpion

Even if you've never seen a scorpion before, you will instantly recognize them by their posture. Scorpions hold their tail up over their back, ready to jab you with their sharp stinger if they feel threatened.

Even though they are venomous, most species of scorpion are not dangerous to humans. Only a few species have venom that is strong enough to kill people.

Young scorpions are born live instead of hatching from eggs. The mom carries her young on her back until their hard **exoskeleton** develops.

Scorpions glow a light blue-green color in the dark when you shine ultraviolet light on them. Scientists don't know how this helps the scorpions, but it sure helps us avoid stepping on one when it is dark!

Great Diving Beetle

Great diving beetles are huge bugs that live in ponds, lakes and slow-moving streams.

They swim to the water's surface and trap air bubbles under their wings. They breathe through slits on their sides, called **spiracles**. To breathe underwater, they suck air from the trapped bubbles through their spiracles.

Larva of the great diving beetle

Great diving beetle larvae are also known as **water tigers** because they are so aggressive.

These beetles are aggressive **predators**. They prey on insects, fish, tadpoles, frogs and even other giant water beetles.

These beetles can fly! At night, adult diving beetles will fly to nearby water sources, using the light of the moon to guide them.

The Publisher: KidsWorld Books

Library and Archives Canada Cataloguing in Publication

Title: Creepy crawlies / Wendy Einstein & Einstein Sisters.
Names: Einstein, Wendy, author. | Einstein Sisters, author.
Identifiers: Canadiana (print) 20190057645 | Canadiana (ebook) 20190057653 | ISBN
    9781988183527 (softcover) | ISBN 9781988183534 (EPUB)
Subjects: LCSH: Insects—Juvenile literature. | LCSH: Insects—Miscellanea—Juve-
nile literature.
Classification: LCC QL467.2 .E46 2019 | DDC j595.7—dc23

Cover Images: Front cover: From *GettyImages:* Parkpoom.
Back cover: From *GettyImages:* EcoPic, nechaev-kon, wichatsurin.

Photo credits: From GettyImages: 15308757, 45; Akchamczuk, 31; AlonsoAguilar, 4; ApisitWilaijit, 42; Atelopus, 59; Ines Carrara, 41; gan chaonan, 52; Cheng_Wei, 55; Connah, 15; CreativeNature_nl, 48; dennisvdw, 46; DeVil79, 54; dimarik, 35; Doctor_J, 33; EcoPic.jpg, 60; EdwardSnow, 14; Leonid Eremeychuk, 6, 7; ErikKarits, 53; FourOaks, 32; frank600, 58; gerard-creamer, 22; JJ Gouin, 36; gprentice, 22; 63; gutaper, 35; Ian_Redding, 5; ironman100, 25; John-Reynolds, 17; bin kontan, 29; Lightwriter1949, 37; maerzkind, 24; marcouliana, 33; Ivan Marjanovic, 19, 20; mccphoto, 27; Motionshooter, 18; nantonov, 31; narinbg, 41; Nataba, 57; nechaev-kon, 56; nikpal, 10, 47; pawopa3336, 49; Photocrea, 12; phototrip, 62; pictorius, 30; prill, 39; reisegraf, 25; richiewato, 26; royaltystockphoto, 50; Tatyana Sanina, 8; scottiebumich, 11; Alexander Seleznyov, 61; selvanegra, 38; SergeyLukianov, 13; SHAWSHANK61, 11; smuay, 16, 54; Srisakorn, 53, 57; Marek Stefunko, 40; TacioPhilip, 51; tskstock, 61; vad_123, 44; VitalisG, 63; vtupinamba, 28; Andrew Waugh, 9; wichatsurin, 43; Adrian Wojcik, 55; wrangel, 47. *From Wikipedia:* Brocken Inaglory, 21.

We acknowledge the financial support of the Government of Canada.
Nous reconnaissons l'appui financier du gouvernement du Canada.

Funded by the Government of Canada
Financé par le gouvernement du Canada | Canadä

PC: 38-1